Tales from the building site

Lessons learned when working on a big site

Pete Wilkinson
www.fastandflawless.co.uk

First published in the United Kingdom in 2020

Copyright © Pete Wilkinson

Pete Wilkinson has asserted his right to be identified as the author of this work in accordance with the Copyright, Designs and Patents Act 1988

All rights reserved. No part of this publication may be reproduced, stored in a retrieval system, or transmitted in any form or by any means, electronic, mechanical, photocopying, recording or otherwise, without prior permission of the copyright owner.

Second Edition

21-07-2020

6 X 9

Dedication

I would like to dedicate this book to all the talented staff in the construction department who I have worked with at Preston College over the years.

You know who you are.

You do an amazing job.

Contents

Preface

Chapter 1 – Introduction

Setting the scene and a few disclaimers.

Chapter 2 – The price work guys, and the day work guys

Different tradesmen are paid in different ways, here we have a look at how each behaves on site.

Chapter 3 – One decorator per room per week

The traditional approach to painting apartments.

Chapter 4 – The finish

The paint finish that you can get spraying is amazing.

Chapter 5 – Sharing the floor with the kitchen fitters

A tale of health and safety in a world where it is possible for Joiners and painters to work in harmony.

Chapter 6 – Grand central corridor

It's as busy as Grand Central Station, trying to spray in a world before self-isolation.

Chapter 7 – Traffic and parking

A decorator's view on a massive UK problem.

Chapter 8 – Theft on site

It happens more than you think, a cautionary tale.

Chapter 9 – Sleeping on the job

We thought that it had to be a carpet fitter, we just needed to catch him out.

Chapter 10 – Playing with the sprayer

The other trades know that they shouldn't do but they just can't help themselves.

Chapter 11 – Coronavirus and Covid-19

It's a pandemic, is it as bad as Spanish Flu? Time will tell, an unfolding tale.

Chapter 12 – Anti reflex

It's a paint from the gods, read about why that is the case.

Chapter 13 – Getting from the bottom to the top of a 20 story building

It more than a story of lifts.

Chapter 14 – Three apartments a week

What can be possible if we get our act together?

Chapter 15 – Out of sequence

Ask yourself, how would YOU make a cup of tea?

Chapter 16 – The masker

They are legends in the spraying world.

Chapter 17 – The dust, the dust, the dust

More dangerous than a hop up but it appears to be less regulated.

Chapter 18 – Tales from the haunted corridor

It's dark and it's long and we are going to try and paint it.

Chapter 19 – It's freezing cold

How we are doing on site in winter compared to our brothers in the States.

Chapter 20 – Delivery a well-oiled machine

Who would think delivering a bit of paint could be that complicated?

Chapter 21 – The challenge of getting water

It's the stuff of life, it's very important to the painter but how hard is it to get hold of on-site?

Chapter 22 – Some final words

Not wanting to end the book talking about water, here are a few parting thoughts.

Preface

A disclaimer before I start.

I have had many experiences over the years while decorating, some of them interesting, some of them inspiring and some of them funny. I have decided to write a book about those experiences for your entertainment.

At the same time, I wanted the stories to be funny and entertaining, but I felt that it was important not to name names and not to point fingers. We all do a difficult job and most of the time we all do our best to deliver the job. I think when things go right it is because people have planned a process and pulled it off.

When things go wrong it is usually because of a breakdown in communication between trades, or even a misunderstanding or the pressure of a looming deadline with its associated financial pressure.

I have collected a bunch of stories and as you read them you may recognise them as something that has happened on a site that you have worked on, some of you may even think the story is about you.

But that is unlikely.

The following stories are not from one job or even from the same decade. They are not based on one builder or decorating company. Some of the stories are true, word for word. Some are completely made up based on things I have seen and felt could have happened. See if you can spot these.

I bet you can't.

I quite fancy writing a piece of fiction one day with a hero and a villain but never got around to it. So weaved into the factual chapters are little stories of fiction.

Some of the stories are true but exaggerated to make them a little more interesting. I have changed names and locations to false ones just to protect the innocent.

My main aim of the book is to make you smile and think about the life we live when working on site.

If you are reading this and you are a decorator yourself, then one of the things I would like to do in the future is publish a book of real life decorator's tales.

Maybe it's a personal journey that you have been on from apprentice to owning your own company. Maybe it is just a funny tale from one of your jobs. If I get enough, I will put them together into a book. You would get a mention of course, a kind of free advert.

If you are not a decorator but you feel that you have a funny story from your workplace then I am still interested, especially if you are one of the other trades.

If you are interested in this, then send me your story to; -

pete@fastandflawless.co.uk

Chapter 1
Introduction

"Building sites" - the very phrase means different things to different people.

The outside world looks in and see a load of high vis jackets and helmets and are a little unsettled by it. What goes on in there they wonder? What are the people on the building site like? But they will never know because large fences are erected, and big signs say, "Keep out".

The public are kept out for a reason, building sites are dangerous and complex things. There are unfinished walls and ceilings, half done electric and plumbing systems. They need to be carefully managed otherwise there is chaos.

Not all tradesman work on building sites but most have had a taste of them during their career. There are many different types of work out there and each one has its own advantages and disadvantages.

Let's have a look at some of them.

Domestic work

These jobs are usually private houses and you will be asked to redecorate one of the rooms or if you are lucky all of them. People usually employ a one man band or a very small decorating firm to do this kind of work, unless of course it's a massive domestic job. A five million pound house for example.

The advantage of this kind of work is that you are usually dealing with the homeowner who is also the customer and the person who is going to pay you. They know what they want, and they will negotiate with you the price and what is expected from you. You will also discuss timescales. These will tend to be realistic.

The work will be interesting, and the customer will want the work to be carried out to a high standard. When the work is completed, they will look at it and give you the thumbs up and then pay you. Immediately, if you have negotiated correctly.

The relationship is simple because it is just you and the customer. No room for things getting lost in translation. If anything changes then it's quick and simple to agree a way forward. A lot of decorators love this kind of work and cannot see themselves doing any other type of work.

There are disadvantages of course. You will tend to take the work on in a piece meal fashion, one job at a time. Not many domestic customers will wait too long for you to start the work so generally you will only be booked up three months at a time.

Some customers can be very hard to deal with and expect you to do unreasonable things like extra work for nothing. Some customers are very slow at paying too.

On the whole though, if you can build a good base of customers then the work can be satisfying and profitable.

Domestic new build

For example, a house extension. These are a new structure that has been built onto the side or the back of an existing house. This kind of work is carried out by a small builder. When it comes to the decorating you may either be working for the builder or the customer direct. This kind of work is like a mini building site but without all the rules and hoardings.

Usually the builder will introduce you to the customer but let you deal directly with them when pricing and carrying out the work. If you work directly for the builder and they do a lot of that kind of work, then you can find yourself with a steady stream of work and you only have to deal with the builder.

The builder knows how you work and how much you charge, and you know that he pays on time and does some good jobs. A side benefit is that the customer may get you back to redecorate some of the existing house for them in the future.

This in some ways is the best of both worlds. You are getting the advantages of working for a builder, but you are only dealing with one person. The work can be interesting, and you can have a steady flow of work if that is what you want.

Commercial

The next kind of work is commercial. Some examples of this are larger refurbishments, these could be a pub redecoration or even apartment refurbishments. Shops and banks are another example of commercial work.

In a lot of ways these are a building site while the work is going on and should follow the same rules as large new builds, but they tend not to. Pub work tends to be done in a short space of time with every trade working on top of each other.

Banks and shops are sometimes refurbished while still open with the work being done in the night while no-one is there.

I have worked on one or two pub refurbishments and I liked them. The building team tend to be the same guys who work together on every job so that even though it's a bit mad, everyone works together in a positive way.

I have found the same with student apartment refurbishment. A fairly small team on the building side with one or two labourers, one site manager and a well-run job. They pay well and on time as long as the work is up to standard and you can have a steady flow of work.

Another branch of the commercial sector is local authority work, and I have worked on this kind of work many years ago when I was an apprentice however, I have no experience of it from a business side so I cannot really talk about it here. Examples would be schools and hospitals.

New build

These can be new houses, new apartments or even new shopping centres. I have had quite a bit of experience of this kind of work and many of the stories in this book come from this environment. I think the sheer size of these jobs cause funny or interesting situations to arise.

What are the advantages of new build?

The first one is a long steady stream of work (in theory) and a large turnover money wise. How much profit comes from that turnover is variable depending on many factors, some of which are out of our control.

The disadvantages are that it can be very difficult to manage a very large job. You are dealing with multiple site managers and many decorators. You will be ordering large amounts of paint and materials. You will have to follow a lot of rules too including having a CSCS card.

The CSCS card forms part of the Construction Skills Certification Scheme. This means that every operative on site has to be qualified and they have to have taken a Health and Safety test and passed. They are then purchase a card which has their photo and qualifications on. Most sites now have a mandatory requirement for trades on site to have this.

Many decorators steer well clear of new build, but just as many decorators make a career out of it and get very good at working on site.

From an individual decorators' point of view, they can make some very good money working for a larger decorating firm on site. The decorators tend to fall into two different camps and each camp has their own set of behaviours that I will look at in the next chapter. There are the price work guys and the day rate guys.

Most of the stories in this book are from the new build environment, so before we start reading, just like you would on the first day on site.

I want you to get yourself a coffee and wait around for the site induction. Then put on your gloves and hard hat and dive into the interesting world of the modern building site.

Chapter 2
The price work guys, and the day rate guys

They are like two completely different breeds of decorator on site.

On the one hand you have the decorators who are "on a price" and they are very focussed. On the other you have the day rate guys who potter along touching up, painting the odd corridor and spending time in the canteen.

Both breeds are working for an employer, neither are truly self-employed although that's the illusion that everyone maintains.

They are both sub-contractors. They work for the decorating company, but they are not "cards in" and that means that they get no holiday pay and no sick pay. They can be "let go" at any time.

From the decorating company's point of view, it means that there is no commitment to find the decorators long term work and no chance of twenty years down the line having to pay them redundancy if it all goes wrong.

From the decorator's point of view, they can do their own work in between doing the sub-contractor work, they can take days off when they want and to some extent they can start and finish the day when they want.

It's the gig economy except we have been doing it in construction for a very long time.

The company will deduct tax at the full rate under the CIS scheme. Currently 20%. So that means that effectively the tax man always owes you money because the tax allowance is not taken into consideration. Many subbies claim their tax back in April and use this as a type of holiday pay or bonus so it can work quite well.

The sub-contractors or subbies as they are known can be either on "day rate" which is an industry standard sum of money that is paid to the decorator, or they can be on a price.

The price is not negotiated, it is given to them. They can choose to accept the price (in my experience most just accept it) or they can go elsewhere and work for another company.

The day rates for trades are pretty much known by the industry and are not really a secret. They vary depending upon market conditions, on the most current job where I have worked, the day rate for the decorator's was £110 per day.

Before tax of course.

Each trade has an expectation of what they want to earn per day, but they don't really vary too much and are between £100 and £200 per day depending on the trade. These days rates start to get a bit set in stone after time and it's hard to negotiate a better deal if you want the steady stream of work that being a subby will give you.

If you are on day rate there is no incentive at all to be more productive unless the boss is constantly on your back asking you to complete things. The day rate guys tend to be slower but do the job right because there is no real reason not to.

Depending on the personality of the day rate person they may choose to spend more time in the canteen than they really should, they may decide to wander around the job and chat with other trades to see what is going on.

They may actually do some work.

In fairness my observations are that the majority are conscientious and get on with the work that has to be done and do a good job. They just don't knock themselves out.

If you see a table of decorators chatting, laughing and joking in the canteen you can bet your bottom dollar that they are the day rate breed of decorator.

Then there are the price work guys. "Don't talk to me I am on a price" one guy said to one of the day raters as he was trying to have a chat.

Ok before we start, what kind of money can these guys earn. Well it depends on the job and the prices that they have been given, it also depends how fast they are of course. Typically, a decorator can earn between £500 and £1000 per week. Some can earn a bit more and some can earn a bit less if they are slow.

So, you can see that there is a real incentive to get your act together and get the work done. The price work guys are on site at six in the morning making a start and, in some cases, still there at six at night. They don't really take breaks because this can cost them money.

If they do have a break, usually lunch then they will bring butties and eat them in the room that they are working on. You rarely see the price work guys in the canteen and if you do it is a brief visit.

There is a big temptation for the price work guys to cut corners, because if they cut corners and do the job quicker then they will do more work and earn more money.

For decorators the typical corners that are cut include, not rubbing down, giving the surface one coat instead of two, using a roller on the woodwork instead of a brush and a few others that will remain a secret.

Some of the price work guys have been doing it for many years and have the whole process off to a tee and there is no way that they would go onto "day rate", once the price work starts to dry up they are off to another site to start the whole price work thing again.

Just to complicate things there are some decorators that do both price and day work. They may work on an apartment all week on a price and then come in on a Saturday to work on a communal area that is being done on day rate.

Even though they are the same person they behave very differently when they come in on the Saturday than they have been all week. They actually change breed to match the behaviour's that we have talked about in this chapter.

Which approach is the best then?

I will give you a minute to pause and think about that one, what would you do? If you are reading this and you are not even in the building trade it may be easier to get a perspective on the situation.

Price of course, more money!

If more money is your thing then of course that's the answer. But don't forget the price work decorators can work longer and harder so that when you divide up the amount that they have earned with the time they have worked it's not always as amazing as it first sounded. But at the end of the day you are going to earn more on price.

Some would never go on price because they find it too stressful and would rather just plod along and earn a predictable income without the pressure.

Money is not everyone's passion believe it or not and some decorators prefer to enjoy their working day and make a really nice job of what they are doing.

Price work can be a bit of game that you can easily get sucked into.

One last note that is worth bearing in mind. Let say that you are a superstar and you manage to do so much work in the week that you earn £1200.

Happy days you would say.

But no, what can happen is that the builder or the boss of the decorating firm sees that you have done well and cuts the prices on the next job. We can have the workers earning too much can we?

This, I feel, is one of the reasons that there is no incentive to be more productive on site and why compared to other countries we are stuck in the dark ages in construction.

The only real way to make more money for yourself is to go directly to the market with your own decorating firm and deal directly with the customers. You are still going to have to be competitive but if you are very productive then you will reap the full benefits of this.

A lot of decorators do not want to take the risk that this involves and prefer to go down the sub-contracting road and in many ways I can understand why.

Chapter 3
One decorator per room per week

I have been a decorator all of my life. For the first ten years I worked for a mid-sized decorating firm and we worked on a range of jobs. From massive churches to schools, hospitals and domestic jobs. In fairness after working for a decorating company I then spent most of my life working in a college teaching painting and decorating to apprentices.

I felt that I had worked on big jobs and I knew what it was all about. Then I got the chance to work on a large apartment building. It was 16 stories high and had hundreds of apartments. I had never been on a job so huge.

There are many of these kinds of jobs in cities up and down the country and the process is a pretty established one. It's pretty established for both the builder and the decorator.

It goes like this.

The rooms are plastered and then allowed to dry. Once dry they are mist coated. This is a thin coat of emulsion painted onto the plaster to seal it. The rooms then have skirtings fitted, kitchens and bathrooms are fitted too along with sockets and lights. Everything except carpet.

At this point the room is ready for decorating and it is given to a decorator. A price work guy. He then does his magic. This is expected to take a week. If the room was given to the decorator on a Tuesday, then it is expected to be handed back the Tuesday after. Nice and predictable.

It does not really matter if the rooms are dotted all over the place and it does not matter if the rooms are not together. It's easy to keep track of and you know who has done which room. The process is fairly slow so that makes it easy to manage.

I am a decorator by trade, but I have specialised in airless spraying and this is the type of work I mainly carry out.

I am used to working on small jobs, some of them refurbishments with new plaster and new wood. The process I use is much faster and I turn the job around more quickly than a traditional brush and roller approach.

I had always wondered what it would be like to spray on a really big job and I also wondered how productive you could be. I was aware of some the problems with builders but not all the issues as I had never done it.

Builders do not really understand spraying (neither do most decorators in fairness) so they base what they know on what they see with the brush and roller guys. If you are spraying, then one of the problems that you can have is that in no time at all every surface has wet paint on it.

When I spray my own work I usually go home once every surface is wet but if you're working on a bigger job then it would be ideal if you could move into another property.

If it was houses that you were spraying, then ideally it would be the next house along so that you are not having to hump your sprayer and associated gear over to the other side of the site.

If you are doing apartments you need all the rooms in a line along the corridor.

I have sprayed many student apartments over the years. These are refurbishments rather than new build, but they are similar. The way that they are done is that we work from the top of the building down. The rooms are stripped out and then we spray them.

I work on a floor at a time and when I am spraying, I move from one room to the next smoothly with no breaks. The sprayer is set up in the middle and I don't have to move it. I can get at the whole floor to spray it if I want to. There is no down time like there is when moving from floor to floor or to the other side of the site.

The main contractor also makes sure that there is no-one on the floor where I am spraying. This is good for two reasons, first of all it means that other trades do not have to wear masks and are not breathing in the paint.

Also, it means that I have no-one in my way when I am spraying, and no-one is going to touch the wet paint once I have finished a room.

It works a treat and I can typically spray a floor containing 15 rooms with one coat on the ceiling and walls in a morning.

When I first arrived at the large new build apartments I was amazed at the sheer size of the whole thing.

Everything took time.

Just walking from the site entrance to the room that you were working in could take 15 minutes. This of course is not production time and is a waste.

You had to carry everything you needed from your van, which was parked off site (more on that later) to the room where you were working. Not too bad if it's just a roller and bucket but not so easy when it's a big sprayer. I will talk about the actual logistics of working on a site in later chapters.

The building was being painted from the bottom up, which was odd to me, but I understand that the building is built from the bottom up so maybe that's why. In the first few weeks I was spraying in a room on the ground floor, then on a corridor on the fifth floor. Then a room on the sixteenth floor. It really messed with my OCD.

Another problem is that on the outside of the building you have a hoist. Which is basically a big lift that takes materials and people to the floor that they are working on.

The hoist flat on every floor does not get finished at the same pace as the rest of the job because people are moving materials out of the hoist and through the hoist flat to the place where they are working.

When the hoist is removed later in the job then these hoist flats need to be finished and decorated. These are not in a line by any means. When they are being painted using the one decorator in one room process it's a doddle but not when spraying.

I quickly found that getting a floor of apartments ready for decorating all at the same time was a real challenge. To this day I do not understand why this cannot be done but I have been told that it's very complicated and that "You just don't understand Pete."

As time has gone on, whole floors have become ready and the new process has kicked in successfully.

Just to simplify the new process puts one decorator in four rooms, ideally two opposite each other on the corridor. The sprayer can be put in the middle of the floor and does not need to be moved when painting.

Once established this new process can deliver more apartments per week per decorator than the old process. It very much depends on the decorator's experience level and it's very much early stages at the moment. I have dedicated a whole chapter to the process later in the book.

As they say, watch this space. I can see all apartments being done this way in the future.

Chapter 4
The finish

One of the reasons that I was recently asked to do some work on some new build apartments was that the builder had heard about spraying. They had heard a couple of things that had gotten them interested enough to look into it.

They had heard that it was faster than brush and roller so of course to them that means cheaper. Another big plus was that the finish was better and that of course means a better standard of apartment for the customer.

It's funny that they didn't equate higher standard of finish to more expensive but then hey it's builders we are talking about here. They are looking to get more for less if they can.

In fairness they did want a superior finish and they were quite forward thinking to consider getting the decorators to spray everything, what we were proposing to do was to spray everything to a finish.

The walls, the ceiling and the woodwork. This has not been done on this scale as far as I know in this country. If it has been done, it's not very widespread.

In the early stages of the job we were starting to spray the woodwork on the corridors, and it caused quite a ripple through the job. Other trades were commenting on the finish and surprised that it could even be done. Some of the decorators thought it was amazing some grumbled that it was better by brush.

A new job had been started and was nearing the point where there was one apartment ready for decorating.

The show apartment.

The boss of the decorating firm I was working for wanted us to spray the show apartment to the proposed finish using the products that we had suggested. The builder would then look at the finished job and make a judgement about whether it was the way forward or not.

No pressure then.

We worked on the room over three days. We worked hard on the preparation and painting to get the room as good as it could be. We were not at full speed ahead because we were discussing the process as we carried it out.

I was timing everything from the masking to the application of emulsion to the spraying of the woodwork. These times would form the basis of the full process.

It was not very efficient because we were only doing one room and there were two of us but that was not the goal of the exercise.

We completed the room and we were pleased with the result. The walls and ceiling looked super smooth, not a roller mark in sight.

The woodwork was obviously done to a spray finish and looked very silky smooth. During the decorating process quite a few people came in to view the room, some of the people I already knew from working on site and some of the people I didn't know and wore suits.

They were all making positive noises and we were confident that we had pulled it off.

Once complete the show flat was furnished with sofas and beds. There were pictures on the walls and occasional rugs.

The room looked very swanky.

From my point of view that was job done and everyone seemed happy. Of course, it had to be viewed by the top guys of the company to give the final approval.

The day came and I got a phone call. "They loved the apartment Pete" I was told.

Phew what a relief, you never know do you? But then there is more. "They said it was the best painting that they had ever seen in their life"

Say no more.

Chapter 5
Sharing the floor with kitchen fitters

This story dates back a few years now but it's still a good one. I was working on some student apartments. These were a complete refurbishment. This included the carpets being ripped out and replaced with new.

The furniture in the rooms was replaced, obviously the decorating was done and also the kitchens were replaced with brand new ones.

Just to give you a feel for how the job is laid out we will look at the size and layout of the building. There were seven floors.

Each floor has three flats.

Each flat has five rooms off a central corridor, each room has an ensuite bathroom. Each corridor has a shared kitchen which is quite a big room. The kitchen doubles up as a common room for the students with comfortable sofas and a large TV.

This means that each floor has 15 bedrooms, 3 corridors and 3 kitchens.

The 3 corridors are linked to a central landing. This central landing is where I set the sprayer up. It allows me to spray the whole floor without moving the sprayer.

It also means that the sprayer does not get full of paint because the central landing and the stairs get done last because these are high traffic areas while the job is getting done.

The kitchen areas are sprayed first because the kitchens need to be fitted fairly early in the refurbishment and it's easier to spray the kitchen once the old kitchen has been taken out but before the new kitchen has been fitted.

The kitchen fitters follow a similar pattern to me, and they work from the top of the building down. The kitchen fitters were a team of three who worked together very well and more or less they could fit three kitchens a day.

Normally I have the whole floor to myself for various reasons so you can imagine my mixed feelings when the kitchen fitters arrived on my floor to start work.

They were a friendly bunch and we had a chat about the possibility of them working on the same floor as me while I was spraying.

The kitchens of course were done so I thought that if they shut the door to the kitchen (all the new kitchen was already in there) and opened a window (it was summer holidays after all) then they would be fine.

So happy with our new work arrangement they got on with the fitting of kitchens and I got on with spraying all the rooms and corridors. We didn't get in each other's way and it was quite good to have someone to chat to while mixing the paint or having a minute between rooms.

Just as a side note to the chapter, the kitchen fitters can earn good money if they are quick and good at their job. It's possible for them to earn £500 per day if they get their act together. I like chatting about earnings and pricing with other trades, it's always interesting to hear what they make. I am well aware that some exaggeration can go on and earnings quoted can sometimes be the highest and not the most regular but it's still interesting.

Many high school career teachers will have told some young aspiring joiner that they should not bother and go to university instead. However, the construction industry has some of the highest earners in the country second only to the medical profession but even then, I am not too sure.

Not wanting to quote actual earnings but I would be confident to say that most construction workers earn more than most careers teachers.

Anyway, back to the main plot.

We continued with this happy working relationship and progressed down the building almost in tandem (not intended) and shared floor after floor as we worked through the job.

Then we had a visit from the Health and Safety guy. He wandered around the job making observations like they do. I was then called in to see him.

In fairness he was a level headed sensible person who was doing his job in a professional way.

"How is the spraying going?" He asked.

"It's going great I said, I have a mask and safety glasses, the rooms are very well ventilated with the windows open and we are using a water based paint."

"I can see that, how do you feel about sharing the floor with the kitchen fitters while you spray?"

I didn't see that coming. I had a think about this. In fairness I didn't really see it as a problem, so I said "I don't think it's a problem, they are in another room, the door is closed, and they have the window open. No overspray will go into their working area."

"What about when they leave the kitchen and walk down the corridor that you are spraying to go to the toilet?" he said.

This was a fair point that I had not considered. Personally, I do not think they would be harmed by a possible 30 second exposure to overspray once in a while. We tended to have breaks and dinners at the same time and toilet breaks are not a regular thing when you are on a price. I didn't say all this though.

"Good point" I said.

He scribbled away on his report and I went on my merry way. The outcome was that the kitchen fitters had to get masks and when they walked out of the kitchen and into the spray area, they had to wear them.

Did they wear them when they did this? Of course, they did.

The year after there were no kitchen fitters on my floor and I was back on my lonesome.

Probably best.

Chapter 6
Grand central corridor

The previous chapter gave you an insight into what it's like being a sprayer on a large site and some of the considerations that have to be made.

You tend to get used to one builder and type of job and when you go onto a different job with a different builder there can be an adjustment period while you get used to them and they get used to you.

Before I tell the tale of the grand central corridor, I just want to tell another tale that gives you more of an insight into what jobs with multiple trades on are like.

This job was the refurbishment of a large detached house. The whole of the inside was being done and most of the walls and ceilings were sprayed.

There were other trades on the job too, joiners, plumbers and electricians mainly.

Health and Safety was a big deal on this job and there was a dedicated Health and Safety person. Let's call him Chris. It's unusual to have this level of scrutiny on a job like this.

Although a big detached house seems like a big job to the man on the street. In the world of construction, it is a tiny job

Chris was keen to meet the decorators, especially when he heard that we would be spraying. He talked to all of us about the job and the safety issues. We all signed risk assessments and method statements so that it proved that we had been told what the procedures were.

The upshot of the meeting was the following rules that I had to follow as the sprayer.

1. Wear a mask, P2 was sufficient. I always wear a higher rated mask than this so that was no problem.
2. No-one must be in the area that you are spraying. That suits me fine, it is just like the student apartment job.
3. No-one must enter the area that has been sprayed for at least an hour after the spraying had stopped. I was not sure about this rule. I don't go back into a

room once it's been sprayed, there is no need for me to, but other trades might, and I am not sure how you would enforce this.

On the whole all the requests were reasonable I thought. So, I decided to spray the largest room first, the lounge. This was massive. I set my sprayer up and closed all the doors to the room and started spraying.

After about 30 seconds the plumber opened the door to the lounge, apologised, "sorry I know I shouldn't but…." And he scuttled though the lounge to another room and was gone.

I smiled to myself knowing that people were not going to stay out of my working area while I was spraying.

Two minutes later the electrician, "sorry mate" and off he went.

While it was not a busy stream of people it was not a "no go zone" either. It did not really bother me or stop me working and in fairness I don't think it did them any long term health damage either, but it shows that keeping an area clear while you spray is easier said than done.

Should I personally have enforced the no fly zone while I was spraying? I cannot stop people from coming through the work area short of locking all the doors. Plus, while I am spraying, I am not that aware of people walking through the area because I am focussed on doing the job.

Time moved on and years passed by and I found myself on a much bigger job with much more going on. Different builder, different job but still the same Health and Safety issues.

I made it known to the people I was working for that it is a Health and Safety issue to have people working in the area where I was spraying and that if we had a visit from the Health and Safety Executive then it would be frowned upon. We put this in our risk assessment and method statement too.

I had a corridor to spray.

It was on the seventh floor and I was spraying the woodwork. This was the door casings (the doors where prefinished) and the skirtings. The corridor was massive, it was sixty metres long, that's nearly 200 foot long for those of you who still think in feet and inches.

There were a lot of door casings and skirting, Because of the length of the corridor I had to move my sprayer and do it a section at a time. I could have put a longer hose on the sprayer, but I decided not to.

The doors were prefinished and had had their door protection removed (for some unfathomable reason) so I had to protect them myself.

There are a number of ways to do this, but I decided to just cover them with "tape and drape" which is a thin plastic sheet with tape on one edge.

Doors are supplied with door protection already on them. It is a plastic sheet bonded to the door. This is designed to be removed once the job is complete. If this has been removed too early, then we have to protect the doors ourselves.

The problem is that the plastic sheet that we use is not as good as the original door protection and can easily be damaged if people are constantly in and out of the rooms.

Of course, that won't be happening because I will be on the floor on my own.

I started to spray the door casings and move down the corridor. This time unlike the last job in the lounge there were other trades coming up the corridor and into the rooms every 30 seconds and constantly. It was the busiest work area I have worked on while spraying.

No-one seemed to mind that they were breathing in satin paint, but they didn't seem to mind damaging the door protection either. In one case they damaged the door protection so bad that I had to do it again.

People also didn't seem to mind damaging the perfectly sprayed door casings either by touching them and opening and closing the doors, this causes dust to get on to the paint.

It was a nightmare to say the least.

On reflection now as I write this, some time has passed and I am a little more experienced, there are a number of things I could have done.

I could have locked the door to the corridor so that no-one could get on the floor even if they wanted to. I could have done the work when there was no one about, for example after 3:00pm in the afternoon or on a Sunday.

I had never experienced so many people on the job while I was working so I was just not prepared for it. I had thought that because I had asked that no-one be there when I spray then that would have been enough.

I didn't realise that a job can be so large that no-one fully knows what is going on at any given time.

In the end, I went back on a weekend and resprayed the woodwork on the corridor. There were still people about but only one or two and they respected what I was trying to do.

Chapter 7
Traffic and parking

If you have only ever worked on domestic jobs decorating peoples lounges who live local to you then you will not be fully prepared for the anxiety and stress that working in another city on a large construction project can cause.

I had some idea because I had worked on student apartments in Salford. The drive to Manchester is known to be bad for traffic but what I didn't know was that the location of that particular job was close to the motorway so not actually that bad to get to and also it was not actually that big so there was enough parking actually on site for everyone working there.

You may be thinking that we all have to put up with traffic in our daily commute and I agree but I think it depends what you are used to.

For me personally I have worked at a college, I lived around the corner and the college had loads of parking. So, traffic and parking have never been an issue.

Once I started decorating again, I worked fairly local and I was able to park next to the job I was working on.

Let's discuss each issue separately.

Traffic first. When I first experienced the Manchester traffic it shocked me. I was doing some teaching at Manchester college, a one day spraying course for the level 3 apprentices. I set off early (or so I thought at the time) which was about 7:00am.

Once I got into Manchester the traffic was crawling. I think it took me an hour to get from the end of the motorway to the college. The whole 40 mile journey from my house took over two hours. I was on time for the class but only just.

Never again.

This is what I told myself, I am never going to drive into Manchester again. Oh, how wrong I was.

An opportunity came up that was too good to miss. It was a large apartment project, it involved some training and some production work, and it involved driving into Manchester. Not the outskirts this time, the centre. I really wanted to do the job for many reasons so I decided that the memory of the traffic must have been worse than it actually was.

Wrong.

I was picked up from my house at 6:20am (the middle of the night as far as I am concerned) and the traffic was every bit as bad as I remembered.

This went on for a number of months. I started making my own way there and found that if I set of at 5:40am (I know) then the traffic was not too bad, still busy though amazingly, but I can do the journey in an hour rather than two. I get there about 6:40am, have some breakfast and start work.

Then at the other end of the day, if you can get away before 3:00pm, say 2:50pm then the traffic is not too bad again. So, this became my routine. I am not a morning person so getting up at five in the morning wipes me out, I know I am a wimp, but I am a "getting older" wimp so that's my excuse.

Many of the trades do this. Some get on the job very early and finish very early. By 3:30 the job is more or less deserted. At first, I was not impressed but now I understand that you have to do this to maintain your sanity.

So, what about parking?

Ok we all know it's difficult to park in a city centre. However, when you are working on a big job that has no parking whatsoever (who cares if the trades can't park?) then it's a problem. What do you do about it then?

First of all, you hope that there is parking near the job that the 200 other people working on the job have not already taken. You hope that the parking fee is not £10 a day because that is £200 a month that you are not using to pay your mortgage.

I have been fairly lucky in that the jobs that I have been working on have had a car park near enough to walk from the car park to the job. The parking fees vary but average about £6 per day.

However, and this is the thing, you need to be at the car park before 7am otherwise you do not get a space. If you have to nip out to get some filler or something, then its game over you have lost your spot.

Another problem with the parking is that your equipment is a long way from where you are parked.

So, you have two choices. You can either hump all your gear from the car park to the job and back again at night or you can leave it on site.

Leaving it onsite leaves you open to theft which I am going to discuss in the next chapter. I have a rule that I never leave my sprayer on site.

So, I have to hump my sprayer (and bag and other stuff) from the car to the job. This is hard work to say the least. Every trade just quietly gets on with it but it's tiring and frustrating.

I recently had a phone call from one of my lecturer friends who wants to go back on the tools because education is so frustrating these days.

I spoke to him about the potential earnings and he was excited, but I found myself trying to talk him out of doing it because of the parking and the traffic. "Try and stay local" I found myself saying "just do domestic work or local site work."

I have concluded that it is not worth it. The traffic and the travel outweigh the money, well for me anyway.

Chapter 8
Theft on site

One of the things that you will notice when you work on site these days is that all trades have a lot of kit, I mean thousands of pounds worth of kit. I cannot decide which trade has the most and which has the least and who spends the most on kit.

Because most sites are massive you have to find a way to hump the kit on and off the job and also hump it around the job as you work, In an ideal world you would work around the job in a logical fashion moving from one property to the next, just moving your kit next door each time.

However, sites are not run like a Borg cube and you are all over the place all the time. Even electricians who you would think would only have a pair of wire snips and a screwdriver have masses of equipment.

Because the kit is so valuable there is a danger of it getting stolen. This is a really bad thing for the individual tradesperson as it means that they cannot do any more work. Usually they have bought all their own tools so that it is a personal loss.

Aside from the financial hit, many of the tools have sentimental value. I have for example a wallpapering brush that I have had all my life and it was given to me by an old craftsman when I was 18. It's probably worth a couple of quid in the pub but to me it's worth a lot more.

Because everyone knows that there is the possibility that their tools could get stolen everyone takes a number of precautions.

These precautions are as follows; -

1. Lock them in a big steel box.

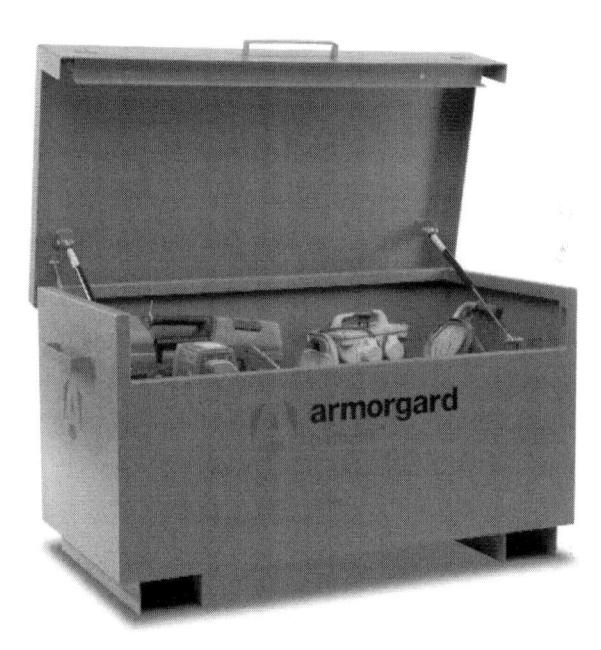

You can either buy these (they are expensive) or rent them. They are pretty big, I mean you could climb inside yourself and they can store a load of tools and materials. Some of them are on wheels so that you can wheel them about.

They are pretty cumbersome, so you don't want to wheel them about much. They are pretty secure too, I think it would be hard for one person to break into one or remove it from site on their own.

2. Lock the item to something. For example, hop ups are quite easy to "borrow" never to be seen again so lots of people lock their hop up to something. The

steel box for example or a part of the building, a steel pillar or steel pipe.

3. Never leave anything on site, this tends to be my approach. If you don't leave it on site, then it can't be stolen can it? The only problem with this is that if you are parked half a mile away from the site then you are humping loads of stuff on and off site every day. This is hard work.

4. Hide it. This sounds like a silly idea, but it actually works really well. If you have hundreds of apartments and each one has a kitchen and bathroom then you can hide something in a bottom drawer, and it would be very difficult to find. I do this with lower value items such as my caulk gun and spray tips.

5. Lock the room that you are in. This is quite a common one. There are not many keys for each apartment. There is usually only 1 key in common circulation, and if you have it then you are pretty sure that no one will get in there.

The next question then is, how much of these precautions are justified. Do things really get regularly stolen off site, bearing in mind that there are CCTV cameras and manned gates and turnstiles that only allow a person through. To get something off site you would have to get past the gate man or past a camera.

To illustrate how possible, it is to steal something I will give an example of something that was actually stolen. Then I will tell you how we think it was done.

Quite a few of the trades on site use airless sprayers. The fireproof coating on the steel is applied by airless sprayer, the plaster is applied by airless sprayer and the paint is applied by airless sprayer. So, you have three trades from three different companies all using similar kit.

One of the common sprayers used on site is the Graco Mark V. This is a beast of a machine. Below is a picture so you can see how big they are.

These cost about £5,000 depending on the specification, so basically the price of a small car. They have a very healthy second hand value also. The plasterers had a few of these machines on site, all of them brand new. They rarely locked them up.

One day one of them disappeared.

Now the only way to get one of these beasts off site was to go through the main gate, past the gate man. So, the gate man was interviewed, no he didn't see anything unusual. The CCTV was checked, no one was seen going out of the gate with the sprayer.

Mystery.

A few weeks later I was talking to the plasterers and they had a theory of how it could have been done. Now I want to stress that this was a theory with no proof, so I am not accusing anyone here. It just makes you think.

The guy that sprayed the intumescent coatings had his own sprayer. It was a Graco 695, very similar to the Mark V.

See photo on the next page.

He was very worried about his machine getting stolen, so he took it off site every day. He wheeled it through the main gate and past the security guy. I have no doubt he even told the security guy that if he saw anyone else wheeling a big blue sprayer out through the gate to alert the management.

The security guy was very used to seeing this one guy wheel out a sprayer.

So, on the day of the theft he waited for a delivery wagon to pull up which blocked half the gate from the security camera. Obviously if he was caught on camera then it would be obvious that the sprayer was a Graco Mark V and not a Graco 695. Hard but possible.

He left early at this point using the wagon for cover and getting away with the Mark V. His own machine he had not brought in that day, but the security guy is different in the morning, so no one noticed this.

A brilliant well executed plan which netted the thief a few grand. The rumour is that he actually wanted the machine to do work that his smaller sprayer was not capable of. He actually told people a couple of weeks later that he had got a machine and was offering the new service.

Nothing however could be proven.

Lots of things get stolen every day from small things like a Stanley knife to boxes of caulk, tubs of paint and boxes of sockets. You name it, if it can be sold in the pub then it's game.

Sad really but some of the staff on site are temporary and low paid so I suppose in their mind they are just making up their wage.

Chapter 9
Sleeping on the job

People travel from far and wide to work on some of the new build work. They make the journey every day and to avoid the traffic they arrive early and if they are on a price they work late. By the time you get home and had your tea it is time to go to bed and start the whole routine again.

London is a good example of this, trades will travel from all over the country to go and work in London. The standard approach is to go down on the train on Monday morning, go straight to the job. Work four long days and then come home Thursday night. If you are on a price and the prices are good you can earn £2K a week which makes it all worthwhile.

While down in London you would have to stay in digs. Maybe a bed and breakfast or a cheap hotel. This adds cost to your job and also means that you have to travel from your digs to the job with your tools. We have already discussed the tools problem so imagine hauling a heavy bag of tools on the tube.

We were working on one job down in London decorating a load of apartments. It was not a big job, there were 85 apartments to do. It was a refurb job so there were other trades there including kitchen fitters, carpet fitters, labourers and of course painters.

One of the carpet fitters, Anthony was always on the job first, no matter how early you arrived. "I like to start at 6" was all he said, "so I can get stuff done before you lot come."

Fair enough I thought, I should try that myself. During the refurb some of the apartments are not done, there is a cycle over a period of 5 years where so many apartments get refurbished every year. This means that the apartments that are left untouched can be used as "welfare" facilities for the building workers. Toilets, showers and kitchen facilities are all available while you work on the job.

I always thought it would save a lot of hassle to just camp in one of these apartments at the end of day. Have a shower, get some tea and sleep on the job. That way you save on accommodation and the hassle of the tube. One or two of us suspected that Anthony was doing this, but we could never find any evidence.

One morning I scouted around the untouched apartment's once everyone had started work to see if I could find anything. Maybe a sleeping bag or some other evidence of a squatter.

The room that we used as our brew room had all kinds of food in the cupboards, cereals, biscuits and crisps. The fridge had milk, butter and bacon in it. There was half a loaf on the side. None of this however was suspicious because we all brought food in for our breakfast, dinner and sometimes tea.

There would always be people there when you arrived and there would be people there when you left. In many cases the same ones worked the long days.

I decided I was going to catch Anthony out and see if I got there early enough, if I could catch him getting out of bed in one of the apartments.

I decided that I would have to get to the job before 6am, officially the job only opened at 6:30am but I knew that one or two of the trades go there before then so I knew that I would be able to get in. The site is actually accessible 24 hours if you have an access code to the gate.

There are security guards on the job but if you get to know them, they are fine, and they cannot keep track of all comings and goings. They are there to keep the public out mostly.

I am not a morning person and the idea of getting to the job before 6am did not really appeal to me but I thought it would be a laugh to catch him out. The next day I rolled up at 5:45am and let myself onto the job.

I thought I would check out the untouched apartments first to see if there was any sign of the wayward carpet fitter.

Nothing.

No signs at all.

I made myself a brew and thought about this. The next thing Anthony comes walking into the brew room with his bag on his shoulder whistling away.

"Morning" he smirked.

I was a but gutted. How could this be? Maybe we were wrong after all. I considered waiting until the end of the day to see if he went off the job, but I had gotten up really early so I couldn't be bothered doing such a long day.

Never mind, maybe it was just my vivid imagination that had gotten the better of me and I had got carried away.

I got chatting with Anthony and asked him where he was staying to see if it was anywhere near our digs, but he was suitably vague saying something about an Airbnb place that he always used that was just round the corner.

"Very cheap" was all he would say.

On the last day of the job he took me to one side and confessed that he had been sleeping on the job but just not every night.

"I do one night in the Airbnb" he confessed. He said he had caught wind that I was going to try and catch him out, so he had gone out early off the job and then come back in again.

Classic, I liked his style and never mentioned it to anyone. I would not be surprised if this is a common occurrence on every site up and down the country it's just that no-one ever mentions it.

Chapter 10
Playing with the sprayer

I am a decorator and when I work on site, I am mainly spraying ceilings and walls with an airless sprayer. Airless sprayers are not a new technology they have been around since the late 50's, mainly in the states but they came over to the UK soon after.

Like most new technologies we have been a bit slow on the uptake as a trade. Decorators can be very set in their ways and are reluctant to let go of the trusty roller. It's funny really because we were slow to let go of the trusty brush and start using rollers in the early seventies.

Back then the roller was seen as a bit DIY and "real" decorators use a brush. Oh, how things have changed, imaging trying to convince a site painter to brush out a new build instead of rolling it, they would think you were mad. Imagine going on site and saying, "Oh you are rolling then are you, is that not a bit DIY?" you would probably get knocked out.

Well sprayers are much the same these days although they are starting to catch on. A roller is faster I am regularly told. A sprayer takes too long to set up and clean is another excuse given.

All this is beside the point except to say that a sprayer on site can sometimes be a talking point with other trades.

Before I tell the tale let me fill you in on some details about the sprayer, I use so that you will understand where I am coming from.

An airless sprayer is basically a pump that pumps paint from the paint tin to the gun at very high pressure. Once at the gun the paint is forced though a tiny hole which is in the "tip" and the tip is in the end of the gun. If you take the tip out, then the gun is just a hose pipe for paint. It's the tip that turns the gun into a sprayer.

The paint at very high pressure and the little tiny hole is what causes the paint to be atomised. It is literally smashed into a mist.

It is a very similar technology to a power washer that you use to clean your car with.

The pressure that you are normally spraying at is 2000 psi (pounds per square inch). This is a very high pressure and to put it in context the pressure in your car tyre is about 30 psi depending on the car.

Everything that is part of the airless sprayer, the fittings, the hose and the gun have been pressure tested to 4 times the pressure that they are rated at. The hose for example is pressure tested up to 12000 psi to make sure it does not explode when being used.

Why am I telling you all this?

Well there are a number of health and safety issues with a sprayer. The first one is obvious and is the fact that you can breathe in the atomised paint. To combat this, you need good ventilation on the job, and you need to wear a proper mask.

The second risk is getting paint into your eyes and to protect against this you need to wear goggles or glasses while spraying.

The final risk is injection injury. This one is less likely to happen but if it does it can kill you, so it is something that as a sprayer you need to know about.

So, what is it? Well the paint is coming out of the tip (a tiny hole) at such high pressure that if your skin is pressed against the tip then you can be injected with paint.

The problem with this is that paint is very toxic and once it is in your bloodstream it is very difficult to get out without surgery. You need to go to the hospital straight away or else the paint will circulate around your system so that it is too late to get out and it will kill you. Within 12 hours in some cases.

Another way that this can happen is if the hose is punctured with a small hole causing paint to squirt out, if you put your hand over the hole then this can cause the same injury. You need to very aware of the dangers and you need to depressurise the sprayer whenever you are changing the tip or adding an accessory.

Not everyone is aware of the dangers of the sprayer and many trades will watch you spraying only to sneak back later to "have a go".

This happened to me, years ago. There were 2 incidents, and both were as worrying to me as the other. The first one was on site and I was spraying out a room and a labourer came in to watch.

"That's great" he commented.

"Yes, it is" I said.

I didn't think anything of it and carried on with the work. When it was lunch time, I left the sprayer and went for my lunch. When I leave the sprayer, I do a number of things to make it safe.

1. Depressurise the sprayer.
2. Put the trigger lock on.
3. Take the tip out.
4. Switch the sprayer off, and I unplug it.

This way it is very difficult for anyone to get up and running and do any damage to themselves and others.

When I came back after lunch the sprayer had been messed with. It was switched on and the gun was in a different place to where I had left it. However, they could not have sprayed with it because I had removed the tip. I suspected the labourer but could never prove it.

The other incident happened when I was training some apprentices. There were 15 in the group and I was showing them how to set up the sprayer and how to spray walls and ceiling. We broke for lunch and we all went our separate ways to get something to eat.

When I returned, I checked all the fittings on the hose and on the gun. I found that they had been loosened by someone. A student I suspected. I tightened up the fittings and carried on with the day. I watched each student to see if I could see who had done the dastardly deed but I had no chance out of 15 people.

The lesson being that you must check your equipment every time that you use it especially if it has been left unattended.

Chapter 11
The Coronavirus (COVID 19)

I could not write a "Tales from the building site" without mentioning this event. I am actually writing this book while in lockdown and I am "self-isolating" the site is still open, but I have decided to keep away for a while. I may edit this chapter down the line when events have panned out, but I will add the edit at the end.

Today is the 25th March 2020, just to give you a timestamp when I start saying things like 2 weeks ago. Saves me having to list a load of dates, this is not a history book.

4 weeks previously I had been on holiday in Paris with my wife and some friends and family. We were there for a long weekend.

We were obviously aware of the Coronavirus and the outbreak in China, I had seen videos of China building a hospital in a week and I remember thinking how amazing that was, not the virus but the fast building.

At this time, I was pretty sceptical of the hype and did not take it very seriously. There are a few reasons for this. Firstly, I noticed that people at the airport where wearing "masks" and these masks could stop dust particles let alone a microscopic virus.

I did a bit of research and found a few startling facts. In the UK on average 17,000 people die every year from seasonal flu. They were predicting 20,000 for Coronavirus. Similar figures I thought.

Globally 10 million people die every year from TB which is the world's deadliest disease.

500,000 people die every year in the UK of various things but usually old age. This is 1,369 people a day. So, when the Government were saying that 50 people (who were vulnerable) per day then that did not seem a lot to me.

This was me in the cynical stage of the crisis, I am sure I was not the only one to be thinking like this.

Then we got back from Paris and we went shopping and noticed that toilet rolls seemed in short supply, for the life of me I still can't get my head around this, we decided to stock pile a bit of wine, although we were the only ones.

Two weeks ago, things started to get strange. Firstly, the roads on my journey to Manchester got quieter. Not really quiet but a bit less traffic on the roads. My routine involved getting to Asda next door to the job and getting a McDonald's (it's a weakness of mine) for breakfast. I noticed that even at the unearthly hour of 6.30 the shop was busy. Not mad busy but busier than usual.

No change on site though except perhaps there were fewer people about.

Then a couple of big events happened. Boris Johnson announced that Schools and Colleges where closing until September and the big one, pubs and restaurants were being forced to close.

Suddenly this was getting serious.

Monday 23rd March while I was on my way to work the roads were dead, I mean hardly anything on the road, it was like being back in 1978 except the cars were more modern. I got to the job and another bombshell, the canteen was closed. People were milling about outside the canteen unsure about what to do.

Then the biggest bombshell of all, and this might be just me, but McDonald's were going to close its restaurants completely. People where queuing for a Big Mac, I mean massive gridlock type queues, not just ten cars at the drive through.

That night while watching the Coronavirus update on the BBC, Boris Johnson announced a "lockdown" in the UK. This meant that you were not allowed to venture out of your house except for a very small list of reasons. People were being encouraged to work from home if they could.

The lockdown period was going to be two weeks. Construction sites were not really mentioned so I could not decide if we were classed as "key workers" or not.

The next day I drove to the job, but this time set off at 7am, I drove to Manchester, got there in 40 minutes which is unheard of at that time. There was no traffic. I got my stuff off the site and drove home an hour later. This time there was actually no cars on the road at one point.

None.

I was like Will Smith in "I am Legend" except I wasn't driving a Ford Mustang and I didn't have a dog in the passenger seat. I was driving to the speed limit as well of course. Oh well nothing like it but in my head, I was looking out for zombies.

I had decided to call it a day and have the recommended 2 weeks off. I got a call from my site foreman telling me that the site was still open and that I could carry on if I wanted but I had decided that it was all a bit too weird for me.

I had also seen a few videos of people (younger than me) with the virus and it looked pretty scary. I like money as much as the next person, but I didn't fancy risking death to earn it.

That night I saw videos of the tube in London packed with construction workers and shook my head.

Should sites stay open?

I will honestly say at this point I don't really know. I think no, at least for 2 or 3 weeks but I suppose time will tell and for you dear reader the next paragraph will tell you. For me at the moment I will have to wait and see.

Here is the update.

I am writing this a month later, we are now in the middle of April. The site did close that night and stayed closed for a couple of weeks.

In that time the senior management put measures in place to make the site a safe place to work. These measures actually filled an A4 sheet of paper but here I will list the most significant ones.

Workers had to maintain a 2 metre distance at all times. Only 1 operative per room was allowed with a notice placed on the door to forbid entry once occupied.

Everyone had their temperature taken on entry to the site and you were not allowed onto site if you had a temperature.

The canteen was completely closed so you had to eat off site. There was a one way route around the job so that people did not cross over on the corridor.

Finally, everyone had to wear a mask at all times while on the job.

People were phased back onto the job and given a new induction to make sure everyone understood the new measures.

I have to say that while sitting in the sun every day appeals to me it was nice to be back at work and to have some kind of routine again. After the experience I am not sure how I would cope in retirement.

The crisis is still not over, and I feel that many decorators have been hit hard especially if they depend on private domestic work. I have spoken to many decorators and they have had thousands of pounds worth of work cancelled on them because people are scared to have anyone in their house.

My view is that many will leave the trade after this is over and go back to employment.

I think that our government have done what they felt was the best approach and it's easy to comment in hindsight. But it is worth considering that not all countries completely closed down their economies.

Sweden have kept pubs and restaurants open and Hong Kong had a vey light touch approach and encouraged people to use common sense, which they did.

I think the crisis has put Boris Johnson in a good light and I feel that people respect his leadership. The NHS has risen to the challenge and I would not swap with any of those people working with infected people in hospital.

They deserve a medal.

Chapter 12
Anti-Reflex

We have been spraying out some new apartments to a finish on both the ceilings and the walls. We have done quite a few and they look great. I was chatting at break time with a new decorator who was originally from Greece. He knew what we were doing but proceeded to tell me that it was impossible for various reasons.

You see conventional wisdom tells us that you can only spray the "mist coat" on new builds, after that you have to apply the emulsion by brush and roller. This is for a number of reasons.

First of all, you cannot touch up a spray finish because it "flashes" this means that you can see where you have touched up. Just think about if you had damage to the door of your car, they would respray the whole door because if they just touched it up in the middle it would show.

Also, the guys that touch up are usually different to the guys that spray so the touch up crew will be using brushes and rollers. At that stage it is very likely that the carpets will be down so to spray would mean a lot of sheeting up. Possible but time consuming.

Secondly when you try and finish using emulsion with a sprayer you can get "banding" this is a stripe effect on the walls that is difficult to avoid. American decorators do something called "back rolling" and this puts a roller texture onto the paint that stops banding and also means that you can touch it up. Of course, "back rolling" sort of defeats the object of spraying, you lose some of the speed and you lose all of the finish.

Thirdly, another reason why it is so difficult to spray to a finish is that if the walls are a colour and the ceiling is white then you have the cutting in problem between the walls and the ceiling. This problem can be overcome of course but it's difficult for most decorators.

All these problems can be overcome because we are currently spraying to a finish so we must have overcome them. However, it was not an overnight journey and it took a few attempts to get it right.

At first, we were in the experimental stage. The job we were working on was using a certain brand of paint (I am not going to mention the brand, it would not be fair) and it was a durable matt. Now even if you're not a painter you will know what durable matt it.

Normal matt emulsion is great to use and looks great, but it marks easily and is not very durable. The paint manufacturers came up with something called "durable matt", all paint companies make this product now because it is very popular.

Durable matt is emulsion with acrylic resin in it to make it more hardwearing. That's fine to make it scrubbable but what it also does is make it difficult to use. You get "picture framing" when you cut in around the edge and then roll the middle. You get banding when you roll the walls and it is very difficult to touch up even with a roller because it has a slight sheen it flashes.

Imagine what durable matt is like to spray. Same problems as a roller but worse. There was no way we were going to be able to spray to a finish with this product.

Another problem we had was that the walls were grey, and the ceilings were white, this meant that we had the "cutting in" problem between the walls and the ceiling. Now this can be done but I think it would be difficult to get lots of decorators doing it quickly because it is quite difficult.

The solution was to do two things, the first was to use a different product and the second was to change the specification so that the ceilings and walls were both white hence solving the cutting in problem with ease.

Which product did we decide to use? Well if you are sharp you will know from the title of the chapter. There is a paint company called Tikkurila who make a matt paint for ceilings called "Anti reflex" this paint is brilliant to spray and leaves no banding, it is also good to touch up because it is so matt.

How good though?

We had a discussion with Tikkurila and asked them if the product could be used on the walls and they said yes that's fine. Then we tried spraying the product in a test apartment.

The apartment was sprayed to a finish and the next day once it had dried it looked awesome. A nice flat matt finish that looked perfect.

The next thing to test is the touching up. I went around the apartment with a mini roller, not a foam one a fluffy one, I wanted to really push it to see how it performed. I rolled here and dabbed there and did quite a bit touching up as a test.

The next day we walked around the room and you could not see any of the touching up, not even in areas where the light cast across it.

Amazing.

The anti-reflex was like magic paint that fulfilled all our desires. We went ahead and used the product on a number of apartments.

Is it that perfect?

Well there was one problem that we encountered once we had been spraying a lot of it, I mean hundreds of litres of it. We found that it clogged the filters up quite quickly.

This is not really a problem because it just means that you have to clean your filters every day, something I do anyway. However, some of the "price" guys saw that as a corner to cut and in some cases did not clean the filters out for weeks.

Not pretty I can tell you. Many filters died in the process of spraying the apartments, split open and destroyed.

So, the lesson is, clean your filters every day.

Chapter 13
Getting from the bottom to the top of a 20 story building

I know what you are thinking, a whole chapter about getting to the top of a building, that cannot be right. Believe me when you're working on a big building it becomes the most important thing in your day.

It can make it or break it.

First of all, let us discuss the order that a building is worked on. If it is a refurbishment and the building is already built then the logical way to work on it, (to me anyway) is from the top down. Working on the top floor is harder than working on the ground floor.

On the top floor you have to get materials up, you have to get tools and equipment up there too. If you need the toilet (these are usually not working in the flats during a refurb) then you have to make your way down to the ground floor.

You are keen at the start of the job and you have more energy for it, plus you are not under as much pressure to get stuff done near the beginning. This is the time to be on the top floor.

As the job progresses and the deadline looms you start to get weary of the job. This is the time to be on the ground floor, easy to get in and out and easy to get materials there.

Typically, on a refurb there are lifts and they work a treat. They have been working for years for the occupants of the building and they are well maintained. The building contractor will protect the lift so that it will not get damaged as the contractors use it.

The lifts can be quite busy, especially at the start of the job when all the work is happening on the higher floors. Old materials like carpets and furniture need to be removed and new carpets and the like need to be taken up.

Sometimes if the lift is being used heavily it is easier to use the stairs. After climbing 10 flights of stairs you appreciate how much easier it is to use the lifts. Even the younger guys are puffing and panting when they get to the top. Even worse if you have to carry paint up, thankfully that does not usually happen.

On the refurbishments I have worked on the paint is very well organised. When the delivery comes, the apprentice unloads the delivery van and then brings the paint up in the lift and loads up each floor with enough paint for the floor. We usually use the kitchen for this. When I get onto a floor to spray it the paint is already there, and I am not wasting my time moving paint about.

I have had a lot of experience working this way and I am comfortable with it.

Then I had an opportunity to work on a new block of flats. This is an entirely different proposition. First of all, and this is a big one, there are no lifts in the building. Well there are but they are being installed and are not available for use.

To remedy this there is a "hoist" on the outside of the building. A large open cage that moves up and down the outside of the building. This is used by everyone to move materials up to where they are working and of course their tools and themselves.

The hoist is a busy lift and is frequently tied up and in use all day by one trade or another. You cannot operate it yourself (for some reason, it's only a lift at the end of the day) so you have to rely on someone being there and being helpful.

Get on the right side of the guy operating the hoist and get organised and it's not too bad.

Another thing that is different with a new building is that it tends to be built from the bottom up so the lower floors are completed first and the trades work up the building. So, when your keen you're at the bottom and when it's rush time and you are fed up, you're at the top.

Now I personally don't understand why the finishing trades such as the decorators and the carpet fitters can't just work top down like we do on the refurbishments. Once the building is built and the roof is on then finish the rooms from the top down. It makes more sense to me anyway and on one job I worked on it was done this way so it must be possible.

So, we are working bottom up, near the end of the job we are on the 20th floor.

Now something that I was not aware was going to happen back then, was that the hoist gets taken out before the end of the job. It's obvious really, I suppose because they cannot finish the building with a big hoist up the side of it and the rooms that the hoist connects to on each floor cannot be finished until the hoist has gone.

Now again if it was me, I would get the lifts working on the inside of the building before I took the hoist down. But no this did not happen. We had no hoist and no lifts, and we were on the 20th floor.

So, before you read this next bit what I want you to do is get a 10 litre tub of paint. Yes, just one. Ideally full of paint, but even if it's half full you will get the idea.

Now carry it up and down your stairs at home 20 times.

Go on, I will wait.

How do you feel?

Well, imagine carrying 2 tubs, 10 times a day. Would you be exhausted?

I told myself it was getting me fit, but I think it was actually going to kill me if I carried on, of heart attack (all those McDonalds maybe didn't help) I mean this is the twenty first century and we are carrying paint up 20 flights of stairs, what is that all about?

I had a bee in my bonnet about if for at least 3 months after the job was done. I have to say that writing about it has made me feel better.

Sometimes we are in the stone age in this country with construction and it's pathetic.

Chapter 14

Three apartments a week

In chapter 3 we discussed how apartments were traditionally painted at a rate of one per week by one decorator by brush and roller. In that chapter I set the scene as to how the old process ran and discussed some of the things that needed to be done to start spraying.

No need to repeat all that here.

What I want to talk about here is something that is not really on the radar of your average decorator. It is something that I think about a lot and it is one of

the reasons that I began to spray most of my work.

Spraying is known to be faster than traditional methods and, in some cases, can be up to 4 times faster. Less on some jobs more on others. How do I know this?

Well I have been on the journey myself and I have found this to be the case. I like to time all the tasks that I do as a decorator and I like to experiment with different ways of doing things to see which is the fastest.

I have learned that it's not just as simple as buying a sprayer, there is more to it than that, it is a change of mindset, and it's one of the things that I teach decorators. Let me give you an example from the real world that shows how much more productive you can be.

I have been lucky enough to work on student apartments over the last five years. We have sprayed most of them and we have done quite a few over that period.

I work for a company and the boss makes the decisions as to what is sprayed and what is brushed and rolled. One of the things that he has decided is quicker to roll are the ensuite bathrooms.

These are small and have lots of masking in them, the showers, the tiles, the sink etc. We didn't do any time analysis, he just "knew" that rolling was quicker. Over the years we have rolled hundreds of these ensuite bathrooms, so we know how long it takes.

The walls are magnolia and the ceilings are white and they are being redone so that they are all white, ceilings and walls. Most of them are done in acrylic eggshell but some in durable matt.

Regardless of which paint we use they take 3 coats by brush and roller for white to cover over the top of magnolia. Each coat takes 20 minutes to cut in and roll.

A couple of years ago, the foreman on my job decided that it would be an idea to try and spray the ensuite and I agreed to give it a go.

I have a chap who masks for me, so I had a chat with him and told him to do the following: -

Don't rush, work at your normal pace.

Time yourself masking a whole floor (14 bathrooms) not a single room. Look at your watch, note the time and then mask a floor, look at your watch again, note the time. He did this and when we divided the time by the 14 rooms, we had the time it takes to mask one bathroom.

6 minutes.

I was surprised it was that quick but obviously that was good news. Now came the time to spray. It took me 1 minute per coat and it only needed 2 coats to cover, so once we had demasked it took 10 minutes a bathroom.

10 minutes!

Don't forget it was taking an hour by brush and roller, so this was 6 times faster. This was in a room that was not ideal to spray and had a lot of masking. Even I was surprised I can tell you.

If you think about that for a minute, imagine if you did all your work 6 times faster, you could complete in a day what you would normally take 6 days to do. It takes a bit of getting your head around at first.

Now I know that in reality it does not pan out exactly like that and while I was working on the job were we were spraying out a number of full apartments it gave me an opportunity to really push the limits of what was possible.

It was a good experiment for a number of reasons. I know that the brush and roller guys did 1 apartment per week, so I had a bench mark. I also had a number of decorators who were going to be spraying apartments so I would get an average of what was possible across the board.

There were a number of decorators working on the apartments and there were a number of variables.

The size of the apartment

Some of the apartments were 1 bedroom, some 2 bedrooms and some 3 bedrooms. So, you can't just say that I did 3 apartments, you have to quantify and say three of the two bedroom apartments.

The length of the day

Some did long days, let's say 6am until 6pm and some worked variable length days. So, to say "per day" must be quantified as let's say an eight hour day.

The length of the week

Some decorators worked Saturday morning, so their "week" was six days, some only worked five days. It is very difficult to get an exact handle on actual productivity.

Single or a team

Some decorators worked alone, and some worked as a team. It may be more efficient to work as a team however for the sake of this I am going to assume that the decorator is working alone.

So, I am going to set some parameters before we go on. I am going assume the following;

1. It's a two bedroom apartment.
2. A day is 8 hours.
3. A week is 5 days.
4. The decorator is working alone.

How many apartments do they do on average in a week?

Well I have spoken to the decorators to get a feel for what they were producing in a week. The consensus is that they can easily complete 2 apartments per week but if they focus and work smart, they can get 3 completed.

I think moving from 1 apartment a week to 3 apartments per week is a massive achievement and just shows that decorators can be very productive if they have the right systems and equipment at their disposal.

Just to put this in context if you worked for Ford making cars and you suggested improving productivity by 10% then that would be a ground breaking change for the industry and you would probably get a £10,000 bonus.

Decorating two rooms a week instead of one room is a 100% increase in productivity. Not just amazing but pretty unbelievable.

What does being more productive actually mean to the decorator? You can use the increased productivity in a number of ways. You can earn more money, you can work shorter days, you can work shorter weeks and you can spend more time on preparation to do a better job. I actually do a little bit of all of the above.

I work a 4 day week and a 6 hour day. I spend more time on preparation and I earn more money than a standard decorator.

Do I believe that we could be more productive? I have thought about this and I think we can. There are a couple of reasons why I think that is the case.

1. The more you work with the new system the faster you will get. I have found that over a period of five years I have doubled my speed

without actually changing anything I have just got better.

2. A lot of time wasters could be eliminated. Things like moving materials around the site. Lack of toilets near where you are working and rooms that are out of sequence.

I did one time study while I was working, and I found that some days I was wasting over 2 hours doing non-productive tasks. That's a lot. Its 23% of the day and there is plenty of room for improvement there.

This is an area where the whole construction industry could improve productivity and is the topic of my next book "Fast and Flawless Systems".

Chapter 15
Out of sequence

I woke up Saturday morning and decided to make myself some toast and a cup of tea. You can't beat a brew first thing in the morning, especially if it's nice and sunny, you can sit out, get some of the morning sun while you drink your cuppa.

I am very particular on how I make my cup of tea, this is how I do it.

I put the tea bag in the cup.
Pour water out of the kettle into the cup.
Then I add milk.

Then I boil the kettle.

Hang on a minute that's not right. It's out of sequence. All the steps are there but in the wrong order. It does not matter though does it? We have carried out all of the steps and the liquid in the cup looks a bit like tea.

Let's taste it. Ughhh! Throw it away. What a waste of time. Make another one but this time boil the kettle before you put the water in the cup. That way the tea will brew.

It's a funny thing but I like things to be done in the right order, maybe it is an OCD thing or maybe I just like things to be right. For example, if I was building a house, I would make it waterproof first and then I would fit out the inside.

There is a reason for that.

If you plasterboard and plaster and then decorate the inside before you put the roof on then if it rains (which in the UK it will do very often) then all your work will be ruined and it will have to be ripped out and started again.

Not only have you wasted your time and materials doing the work the first time round, also you are a little bit pissed off and you probably will not do the job as well the second time round.

If you are not a builder and you are reading this, you probably think that I have gone a little mad and this is one of my made up stories because it can't possibly be true. But guess what guys? These things go on.

I find it so frustrating when I see it happening, I have to close my mind to it to cope. How can anyone be so stupid? I am told that I don't understand and that a building is a complex thing. There are many factors to consider.

Mmm, I am not too sure.

Another example that more directly affects me as a decorator are the two trades either side of me. By that I mean that before I go into a room it has to be plastered and the skirtings and architraves are fitted. Then we decorate and then the carpet fitter comes and lays the carpet.

1. Plaster
2. Fit woodwork
3. Decorate
4. Fit the carpet

Just look at that list for a moment and consider if you were doing work in your own house. Your own house is no different to a building site really, it's just the scale that changes. You have one house, the builder has a hundred.

This does not affect the order of the work.

So, you would not for example fit the woodwork then plaster, would you?

Oh, hang on a minute, that has happened on jobs that I have been on. Plaster all over the woodwork, no edge to the skirting because the plaster has levelled the edge off.

What!?

Another funny thing is that if you were doing work in your own house then you would probably do a plan on the back of a fag packet.

Building companies have complex and expensive software that they use that plans the whole job. They create programs of work which all the trades are supposed to adhere to. They have teams of people in the office keeping all this up to date. They wander around with iPads so that they can keep all the progress up to the minute in real time.

Wow, why then is the work not done in sequence?

My favourite one, or least favourite really are the two following situations.

Firstly, the carpet fitters fit the gripper and underlay before the decorator goes in the room. The logic behind this is that it progresses the job quicker. Sorry I don't see how but there you go.

No big deal you would think, the carpet is not down, you can just crack on and paint. But think about it. You cannot clean out the apartment properly because there is gripper close to the wall. The dust is piled up between the gripper and the wall. It's very difficult to hoover it out.

It's very difficult to even get a hoover on site.

If you are spraying, then all that dust blows up all over the wet paint and looks terrible. A bit like pouring cold water on a tea bag, the result is horrible.

Plus, when you rub the woodwork down with sandpaper you rip all your fingers to shreds on the gripper. Ouch!

The underlay gets torn to shreds too but that is just something that pisses the carpet fitter off.

Secondly, the carpet fitter actually fits the carpet before the room is decorated. So now we can't spray it easily and end up rolling it which takes twice as long and looks rubbish. The carpet can get damaged because we are working on it too.

In some cases the carpet is fitted, then they realise the room is not decorated (how can you not notice that the room has not been decorated?) then they roll the carpet back from the edges and fold it up in the middle and chuck a sheet over it.

Great!

I could go on with my out of sequence tales, but I won't, I think you get the idea. So, when you are doing work on your own house and it all goes to plan. You may think that you could do it as a job and run a bigger site.

Well I would say that you could not do any worse than some of the professionals.

Chapter 16
The Masker

One of the things that I am interested in as a decorator is working more efficiently. When I tell people that I do a lot of spraying their first reply is to talk about the masking.

"I bet the masking takes ages" they say.

This is a very common thing that is said to me, it's funny really because sometimes I think that they think it is the first time it has been said to me and I have been spraying all this time without realising that I am wasting all that time masking.

I have given this much thought and there must be a reason why decorators or people in general think that this is the case. Let me explain what I think happens.

You decide that you are going to spray something, let's say a room. They are a little apprehensive because they have not sprayed before. They have not masked much before either so off they go and start masking.

It takes ages.

When they have finished, they look back on the time that they have spent masking and they think to themselves, wow masking is slow.

Now this is interesting because for some reason they don't say to themselves, "Oh I am a bit slow at this" they assume that they are good at it and it is the process that is slow.

Imagine if you applied the same thinking to walking. If you have kids, you will have watched them learn to walk. If you haven't you will still know what I mean.

They want to get about the room and are a bit sick of just sitting there. Their legs are not really strong enough to support them yet, so they do the next best thing, they crawl. Boy it is slow, but it works, and it gets them about.

Then their legs get stronger from all the crawling and they haul themselves up on a chair or something so that they are stood upright. Then they take their first steps, it's a bit wobbly and it ends in a fall.

Do they give up on walking?

No, they don't because they can see us doing it and they know it can be done and it can be done well so they work at it.

Masking is the same, you get good at it. Check out videos from "The Idaho Painter" on YouTube and you will see how fast you can get with masking, in fact I think that it is faster to mask a line than "cut in" a line and you only have to mask once.

I have been training decorators to spray apartments on a large job. Some of these guys have never sprayed but they have painted many apartments over the years with brushes and rollers, so they know how long that takes.

The first thing I do is I show them a room that has been masked by someone else. I point out all the little details that need to be thought about.

Then they have a go. It takes a while at first, but they do a room, it usually takes them about 5 hours. They shake their head and mumble about not being sure that it is the quickest way. I tell them that will get quicker, but they are unsure.

Fast forward a couple of weeks and they are flying, in fact they are actually enjoying it. They say that they would not mind masking all the time. However, that is not the system we are using, and they will get to spray as well and that of course is the fun bit.

Which is the more skilled task? Spraying or masking. I cannot decide myself and I have done loads of both. I think though that actually masking is underestimated, and people think that there is no skill involved and that its easy but in actual fact it's a real skill to be mastered.

On another job that I work on we have a chap who is a masker. That is all he does. When we spray the student apartments, I spray them a floor at a time, one coat on both the walls and ceilings.

I work from the top of the building down (as you know) I usually do 2 floors a week to a finish. The masker then has to get ahead of me, he will mask a floor and then hand it over to me to spray. I like to let him get a couple of floors ahead of me before I start.

There is quite a bit of masking to do but not loads. There are the sockets and switches. The desk and wardrobe. The doors are prefinished, so we mask them off too.

He does it day in and day out so he is really fast at it, faster than I would be probably. I check the rooms before I spray them just in case, he has missed something.

There is always the odd thing and it's better to catch these before I start spraying otherwise you will end up getting paint all over something that you shouldn't, and it destroys your flow.

The system works really well, and I think that a dedicated masker could be the most efficient way to work.

Chapter 17
The dust, the dust, the dust

You would think that a hop up would be a pretty straight forward thing. I mean it's just something to stand on so that you can reach that little bit higher.

In the old days we used to use a milk crate, they were perfect, light and strong and best of all free. You could nail two to a plank and you had a makeshift scaffold.

I have a couple of hop ups that I use in my decorating business, one is 300mm by 600mm and is quite sturdy and useful.

I also have a little tiny hop up that is really handy and easy to carry, like the one below.

The really small hop up is the one I used the most. Can you imagine my surprise when I discovered that neither of my hop ups are allowed on site and that you need one that is 600mm by 600mm?

This I am told is a Health and Safety decision. Let me tell you these massive square hop ups are not easy to use, especially in a small space like a bathroom. You end up banging and clashing against everything.

Now I am all for Health and Safety measures if they make sense, like wearing a mask when spraying but what difference does it make if your hop up is 600 by 300 instead of 600 square. I have never fallen off either of my hop ups and even if I did, I don't think it would kill me.

Dust however does.

Not straight away of course but over time, most dust on building sites is silica dust, and it will kill you eventually if you breath it in. Now I am not having a go at any one site here because I have worked on many sites over the years and they are all the same.

There is dust everywhere, you really notice it as a decorator because we are the only trade that actually tries to get rid of the dust.

The official line is that you have to take 3 main measures to eliminate the hazard.

1. Wear a dust mask. A proper particle one like the one below.

2. Use a dustless sanding system when sanding down walls, ceilings and woodwork.

An extractor

A wall sander

3. Use a proper hoover to clean up dust. Do not use a brush.

Bearing in mind how strongly the hop up rule is enforced (they throw them in the skip if they are not big and square) can you imagine my surprise when I saw all the walls being sanded by hand. With mountains of dust building up on all the skirtings in every room and all the corridors.

I am talking dust in quantities never heard off, dust everywhere. Do the sanding people clean up this dust? No, it is left for us to deal with.

Now if we are lucky the labourer comes along and sweeps it all up into a pile. It's like an episode of stars in their eyes when this is going on. But hey they are wearing a mask so It's ok, not sure about everyone else though.

What I can't fully understand is that it would be quicker to sand the walls with a proper wall sander with an extractor, it would do a better job and it would save all that time sweeping up.

Last but not least it would save all that dust from being breathed in by everyone. Thus, saving lives.

Chapter 18
Tales from the haunted corridor

Winter on a building site is cold and it is dark. The inside of the building, the corridors mainly need to have temporary lighting so that you can walk around the site safely. The rooms however do not have any lighting and rely on daylight to work by.

The site electric is supplied from a generator up until the main electricity is connected. You cannot really tell the difference much when you are in the building except that sometimes the lights pulse and go out completely when the generator runs out of diesel.

During the winter period most of the trades will have their own site lights that run off 110v so that when they are working in the rooms they can start early, before it gets light and work later once is goes dark again.

These lights come in various shapes and sizes from large lamps on a stand that illuminate the whole room with ease to small battery lamps which clip onto your helmet. Many trades such as the electricians need their light focussed on a small area while they are working so a small lamp is plenty for them.

Plasterers and decorators need the whole room lit up so that they can see what they are doing. It's hard to work under artificial light and get a good finish.

The rooms can be worked on through the winter and then when spring comes, and the sun comes out and casts it's all seeing eye across what you have produced it can look a bit ropey. It looked fine in January under the site lamp you say.

The corridors tend to be used by everyone and therefore they are lit permanently with site lights provided by the builders. These light the corridors really well and mean that you can see what you are doing even in the depths of winter.

We were working in one of the rooms off the corridor on the fourth floor. What you need to bear in mind (and this is difficult to visualise) is that the corridors are very long. 60 metres (200 feet) long in many cases. That's a lot of lights needed to illuminate the corridor properly.

It was first thing in the morning, and I had set up to work in my apartment. A little light set up so that I could see. This room was at the end of the long corridor.

I needed to get something, opened the door to a pitch black corridor, with scratching and moaning noises coming from the blackness.

I froze for a moment. Oh shit. I can't see a thing.

Then this voice came from the depths of darkness. "Sorry mate I have been asked to take the lights out, just doing what I am told, sorry again."

I paused for a second, the fear was starting to subside, and it was replaced by anger. I was about to tell the guy that this was unbelievable, and could he not at least leave the lights in until we had finished the room.

Then I considered that the poor bloke was really just doing his job. You see at some point in the building program the lights are scheduled to come out and the actual corridor lights that have been fitted will be switched on.

Now call me old fashioned but I would fit the actual corridor lights, then switch them on and then take the temporary lights out. The poor bloke taking the lights out was working in the pitch black (that can't be safe) and struggling.

At this point I switched on my phone light and navigated down the corridor. It was a bit grim, I had a quick conversation with the light removal bloke and went down a floor to get something from one of the other decorating teams.

On my way down I bumped into James, one of the decorators.

"What are you up to James?" I asked.

"I am going up to four to spray the corridor" he replied.

It was at this point for some reason I started to see the whole situation as bloody hilarious. Though I hid that for a second, I wanted to see what James thought.

We went up to the corridor on floor four and it began to dawn on James what was happening. Bear in mind James is on a price.

"Oh my god" he said, "this is unbelievable, how I am supposed to work in these conditions, I can't see a bloody thing."

At this point I just collapsed laughing. The situation was so ridiculous that I could not take it seriously.

"I am going to write a book about all this, it is hilarious" I said.

I think James started to see the funny side too. We got a couple of our own site lights and set them up in the corridor. These were a little bit under powered really and only lit part way down the corridor, but your eyes adjusted, and you got used to it.

James set up his sprayer and started to spray in the dark, once we had gotten over it, we just got on with it. I wanted to take a video of it all, but it was just too dark.

All I can say is wait until it goes dark and then switch off all the lights in your house. Get yourself a torch and imagine decorating your house under those conditions and then you might have a small insight into the challenge that we faced that day.

Chapter 19
It is freezing cold

In the previous chapter we discussed one of the problems with working on a site in winter, the dark. This problem is fairly easily and cheaply remedied with a few site lamps. However, there is another fairly obvious problem and that is the cold.

Now building site workers are a macho lot, when people say to them that it must be cold in the winter and you hear tales of bricklayers breaking the ice off the water in their water bucket they should laugh it off.

Each trade will have different strategies to cope with the bitter cold. If you work outside, then you will have several layers of clothes and socks. You will have gloves and a woolly hat.

Obviously in some cases it can get too cold to even work outside, depending on the job that you do. So if you are a plasterer and you also render the outside of buildings then this cannot be done in winter. There is a render season that is in the warmer months and this is when the bulk of the work gets done.

If you are working inside a building, then in theory it should be better. You are shielded from the wind, so this means no "chill factor" also you are at least dry.

For some trades, for example the electricians who more or less always work inside, a small heater next to where they are working can be enough to keep them warm, once the wiring job is done then they can move onto the next job and the cold is not going to affect what they have just produced.

The fact that modern building is well insulated can be a bit of a double edged sword because what it means is that if you do have a smaller heater then the area can warm up fairly quickly. If you don't then sometimes the inside of the building can be colder than the outside.

Yes, colder.

For the painters the cold can bring a whole raft of problems. In summer the paint will dry quickly. In some cases, the paint can be sanded and recoated in a couple of hours. If you are on a price it is important that the paint dries so that you can move onto the next stage and complete the work.

In winter water based paint just does not dry. This is not always understood by the other trades or the site managers, they just think that the paint dries slowly. They wonder what the painter is moaning about. But once the temperature drops below a certain point, around about 4 degrees then the paint will not dry at all.

Some of the decorators painted their woodwork on Friday to allow it all weekend to dry only to find that on Monday it was still wet.

You may be reading this and thinking, why don't you just get a fan heater or something? But what you have to remember is that one fan heater will not really heat a whole apartment or house, plus the fan heater would have to be 110V or it would not be allowed on site. You can't just bring one from home.

So, you may need a couple of heaters to heat an apartment. Then you may have 10 decorators working in 2 apartments each. So that is 20 apartments, 2 heaters per apartments so that 40 fan heaters.

Unless you bolted them to the floor, they would go walkabout or in simple terms get stolen. Plus is it really safe to leave a fan heater running all night? I am not sure that is a good idea.

I have my own training academy and we teach spraying to decorators all over the country. Because of this I have a large workshop area. This area is quite difficult to heat. We manage during the winter months by having a number of electric heaters on during the day, but it is never toasty warm.

One year we had a guest over from America, Idaho to be precise. He was a chap called Chris Berry and he has a YouTube channel promoting (among other things) spraying.

He was over in the UK to spend some time with us and see how we do it over here. He was due to deliver a training course at our centre.

It was late November.

In fairness the weather was good, it was sunny and dry every day, he actually thought it was like this all the time.

In Idaho, they have a beautiful climate and it is warm and sunny all summer.

Because the weather was sunny that meant that first thing in the morning it was also cold. Chris wanted to go up to the workshop the day before to check it out and get the lay of the land.

He walked into the workshop and immediately commented on the cold.

"It's quite cold in here" he said.

So, me being me just replied "Oh Chris, have you never worked on a building site?"

I got a blank and confused look back from him.

"You know freezing cold in winter."

He shook his head and went on to explain that on building sites in America they put in a temporary hot air heating system while the building is being built. That way the workers are working in a pleasant environment and the dry lining dries, the paint dries the water pipes don't freeze. It all seemed pretty obvious to him.

At that point I realised that the Americans must look at us as if we are in the stone age.

"Why would you not have temporary heating? You cannot build a house in the freezing cold" Chris said.

No Chris you can't, but we try our best. Sometimes I think we need reminding in this country that we are now in the twenty first century, not the nineteenth.

Chapter 20
Material delivery

When you are decorating your own home, it is pretty easy to get materials. You pop down to your local B&Q in the car, pick up a couple of tins of paint and then you are sorted.

When the jobs get bigger then there is little more planning involved. For example, if I were to decorate a whole house I would measure up the job to work out how much paint that I needed. If there were any wallpapering to do then the customer would have to choose this and then very likely I would have to order it.

I would probably get the paint delivered to the job while I was there. It would not really matter what time the delivery van came because I would just look out for them. They could pull onto the drive and just unload the van with ease.

If the job is bigger and it has other trades, then it gets more complicated. A house extension costing say £30,000 would have a few trades on the job. There would be bricklayers, joiners, groundworkers, and then later on there would be electricians, plumbers and painters.

There would be more than one delivery a week and the materials would need to be stored somewhere while the work is carried out. Some of the materials would have to be stored in a dry place.

There would only be relatively small amounts delivered though. Just enough bricks and blocks for 1 extension. One roof, roof tiles, maybe ten door sets and a few windows. Then later, 1 kitchen, one boiler and a few radiators, some sockets and light's and maybe a consumer unit.

There would be nothing that you could not store outside with a sheet on it or in the garage.

On these types of jobs, you do not really think about delivery of materials and you do not worry about running out of materials either. If the worse happens you could always nip to the supplier in your van and pick up some bits and bobs.

What about an eighty million pounds apartment building though? All of a sudden you are dealing with larger quantities of everything. For example, if you had let's say 400 apartments and each apartment had 10 doors then that's 4000 door and casing sets. Imagine that in your garage!

There would be 400 kitchens and 400 boilers. That's a lot of expensive stuff that needs to be stored safely and managed on site.

Just looking at the problem from a painting point of view. If you decorated your lounge, then you would probably buy a tin of paint. This would very likely be a 5 litre tin. Most decorators don't use 5 litre tins because it does not cost much more to buy a 10 litre tub.

On a larger job you are looking to order a pallet of paint. This varies from different paint companies, but a pallet usually holds about 48 x 10 litre tubs of paint. That's over a grand's worth of paint. One of the jobs had one of these delivered every week. That's how fast the paint was being used.

Running out of paint is out of the question, you can't have people standing about waiting for paint and nipping to the paint supplier to pop a pallet of paint in the back of your van, it is out of the question, they just would not have that amount of paint in stock.

Then you need somewhere to store it that will not get too cold. Frost will ruin emulsion paint, so the paint needs to be kept undercover.

The next problem is delivery.

Usually big apartment jobs are in the middle of a city, London for example. There is very little parking around the job and the gate to the job will have limited space for lorries to pull in and make a delivery.

So, what does this all mean? It means that deliveries of materials become quite an operation that needs organising very well. Each company on site will have a delivery slot of two hours and they will negotiate with their own supplier so that the stuff is delivered on time.

For example, the paint delivery slot may be Monday between 10:00am and 12:00am. The delivery van will pull in at this time and the paint would be unloaded by forklift. The paint is then transported to the allocated storage area.

If the paint supplier misses the slot for some reason (traffic for example) then they will have to come back some other time. This would cause a problem though because it means that the job could run out of paint.

This process was very well managed on the site that I was on and lorries are coming in all the time, being unloaded and then off they go. A steady stream of deliveries all day.

Then it is all stored in the basement, which is massive and compartmented off into locked stores for the different trades. An amazing array of boilers and kitchens and door sets and pipes, it's like B&Q in the basement of the building.

Chapter 21

The challenge of getting water

If you are sat reading this and you decide that you want to have a glass of water, then it is a fairly simple process. Get up, walk over to the sink with a glass, turn on the tap and get your drink.

Simple.

If you are a decorator on a job, then water is pretty essential, and you will need quite a bit of it all through the day. Especially if you are spraying.

The paint will need a certain amount of thinning, a lot of thinning if you are mist coating. I would typically spray about 200 litres of paint in a day, so that is 20 x 10 litre tubs. If I thin by 20% (it varies depending on the product I am using) then I would need 40 litres of water. That's quite a lot.

On a small job it's a simple process, most people have a utility room or an outside tap. One job I worked on had an outside Belfast sink with hot and cold water.

Luxury.

On a large job that is being built there is a temporary water supply. This may be a pipe running up the stair well with a tap on each floor or it may be an outside tap somewhere on the site.

It is usually a blue pipe coming out of the ground with a plastic tap on the end, nothing like the picture above which is an ideal world tap.

The water situation is a little bit like the light's situation only more labour intensive. So, the water gets taken out at a certain point in the job and you could hope that at that point the water was switched on in the properties but that is not always the case.

Once the water is taken out of the building then you have to get your bucket and make your way down the stairs (remember you could be on the 20th floor) and fill your bucket with water and then carry it up to where you are working.

This feels like a complete waste of time, each journey can take 20 minutes and don't forget you will need 4 or 5 of these buckets of water a day.

When you come to wash out the sprayer then the ideal is to use warm water because this will clean the sprayer out much faster than cold.

Well you can imagine how difficult it is to get hold of warm water if cold water is such a challenge.

What happens?

Most guys just don't bother to wash the sprayer out and it will get cleaned if it is lucky once a fortnight. This will cause all kinds of problems, especially with the filters.

The brush and roller guys are not bothered as much by all this. They will not thin the paint as much and may even use it neat and they will just "bag up" there rollers so that they stay wet for the next day.

Bagging a roller just means putting the wet roller into a plastic bag so that it does not dry. A roller can be kept like this for weeks.

Most decorators will take the rollers home at the weekend and wash them out. An old washing machine in the garage is the favourite way to do this.

For me I think the water issue was the most tiring and time consuming. This really is a "productivity" issue that we looked at in an earlier chapter.

I recently worked on a domestic job and had a break from the site work. To have the water (hot and cold) at hand and a toilet and a kitchen was like I had died and gone to heaven.

Chapter 22
Some final words

I didn't want to end the book moaning about how hard it can be to get water on site. Although after reading the last chapter you may think that I just wanted to have a good moan about work. I hope that is not how it has come across.

I have written this book for both people in the trade and also for people that have jobs on other walks of life and wonder what it is like to work on a large building site.

If you work on a building site yourself then I am sure that you recognise some of the situations that I have talked about and hopefully they have made you smile.

If you are another trade and you have a story to tell from your perspective then I would love you to email me a rough draft of your story at

pete@fastandflawless.co.uk

I may write a Tales from the building site 2 if I get enough stories.

I wanted to include stories that were eye opening and maybe a little shocking, but I also wanted to inject a bit of humour into the stories. I also wanted to look at some of the positive aspects of working on a large building site.

We live in changing times to say the least. When you look at the jobs that are in the most demand now you will probably find that they did not exist ten or twenty years ago.

The white collar jobs are in danger of being replaced by AI or artificial intelligence. AI is progressing at a very fast pace. I have an Amazon Echo at home which is a sound device that is connected to the internet. On the server Amazon have an Artificial Intelligence software running. This is being constantly updated so that the Echo gets more intelligent every day.

Many retail jobs are being replaced by online shopping and the trend looks like it is not going to be reversed.

For the past twenty years the younger generation have been encouraged to go to university and get a degree, these days even if you are not that clever. Over 50% of the population have been to university and there are only really 5% of jobs in the economy that need a degree.

The Government has sold the degree route with the promise that you will earn more money over your lifetime if you have a degree. This has shown to be a lie and, in some cases, you could actually end up earning less than you would have if you had just gone and got a job.

On top of that many graduates will come out of university with a massive debt that they will not be able to pay off over their lifetime. The debt is written off after 30 years and after that the taxpayer picks up the tab. Degrees are overpriced and in some cases are worthless. We are not getting a good deal as taxpayers or graduates.

There is an alternative that your career counsellor at school did not tell you about. You could have a career in construction.

I was speaking to a chap (let's call him Frank to protect his identity) who was leading an association in the decorating industry. He told me a great story and I think it sums up the whole situation.

When he was at school his teacher gave him advice on what to do in the future. Go to university and get a degree and then get a good job.

"I quite fancy being a joiner" Frank said to his teacher.

"Oh no, you don't want to work in construction!" was the reply. "Luke who is a bit of tearaway is going to work in construction because he has no choice."

Frank went to University and took on a load of debt. When he had graduated, he struggled to get a "graduate" job. One afternoon while walking out of Asda he saw the legendary "Luke" from school who had got himself a job as a joiner.

He was driving a Porsche.

Right there Frank realised that he had took the wrong path and he had been lied to.

Luke waved and had a chat. He was really busy with work as a joiner, he was in demand and he was earning great money. He also got a lot of job satisfaction from what he did as well.

Life was good for him.

This has been my observation too. If you learn a trade, then you can have an interesting job that can take you all over the world. You will always be in demand and you will earn good money. Plus, you will have a varied and interesting job if that is what you are looking for.

If you want to really make a difference and earn big bucks, then you could train as a construction professional and then you could make sure that all the problems that I have outlined in the book will never happen.

We need people like you.

I worked with a chap at college called Steve. He had worked on Wembley Stadium and was an expert in concrete. He got a job teaching at college and I worked with him for a while.

He took a pay cut to work at college. I think the college paid him £26K. I could not for the life of me understand why he had come to work at college, he was very good at his job.

He told me that teaching was something that he had always fancied doing and that he wanted to give something back to the industry.

The students liked his classes and he took us all on a few trips, we went to a quarry to see how they quarry stone to make concrete and we went to a concrete batching plant. There is more to concrete than meets the eye.

Anyway, after a couple of years his old boss contacted him and made an offer to tempt him back into industry.

He decided to take it.

I asked him what the offer was, and he would not tell me, all he would say that it was "in excess of six figures."

So, he took his £100,000 plus and went back into the construction industry. Shame because we lost a good teacher but in fairness, I don't blame him.

With that thought in mind I am going to have an early dart and call it a day. Maybe one day I will see you working on site.

Take care.

Other books by the author

Fast and Flawless

A guide to airless spraying

This is a chatty guide to airless spraying for decorators, decorating students and anyone interested in spraying with an airless system. The book covers all aspects of the airless sprayer including the component parts of the system, the different systems that are out there to buy and setting up the system.

The book covers topics such as types of sprayers, essential equipment, using the equipment, masking, PPE and masks, a bit about paint, what to do when it all goes wrong, spraying in the real world and common paint defects.

Fast and Flawless Pricing

A guide to pricing and business for decorators

Are you a decorator that struggles with pricing?

Have you just set up in business and are looking for some pointers?

Are you an established business looking for some inspiration on how to move forward?

This chatty guide on pricing and business will gently guide you through the process of pricing a decorating job. It looks at the pitfalls of getting your pricing wrong and the advantages of having a good pricing system.

The book has been written by someone who has both been a decorator and taught decorating in a local college for most of his life.

Fast and Flawless Systems

A Decorators guide to planning and carrying out successful job

This book looks at systems for Decorators.

This book covers all types of systems from which paint to use on what surface to what order you should spray a room. The book also covers aspects of decorating that you may or may not be aware of such as painting uPvc, training, funding and marketing.

If you have read the other two books already then this is one is a must read, if you haven't then this book is a great place to start.

Boat Life

The trials and tribulations of living aboard

Nothing to do with construction or decorating. I love boats, I have one and I have lived aboard myself, so this is an insight into the lifestyle.

This is a book for boaters, written by a boater. Pete Wilkinson has spent his whole life around boats and has owned a couple too.

The book looks at all aspects of boating including, what is the best boat to buy, where to look when buying a boat and do you build one or do you buy one?

The following questions are answered; - Which is the best type of boat - Wood fibreglass or steel? Do you borrow money or save to get your boat? Do you move around or stay put? What is the essential kit needed? How do you keep her shipshape?

The book also looks at living board and gives an insight into the liveaboard life. The book also puts boating into the context of modern life and discusses the advantages of living on board.

Finally, if you have wondered what goes on behind the curtain of a boater's life then this book will show you.

Boat Life

Pete Wilkinson

The trials and tribulations of living aboard a boat

This is a book for boaters, written by a Boater.

Pete Wilkinson has spent his whole life around boats and has owned a couple too.

The following questions are answered:

1. Which is the best type of boat to buy: wood, fibreglass or steel?
2. Do you borrow money or save to get your boat?
3. Do you move around or stay put?
4. What is the essential kit needed?
5. How do you keep her shipshape?

The book also looks at living aboard and gives an insight into the liveaboard life. Finally, boating is put into the context of modern life and discusses the advantages of having a boat in your life.

Pete Wilkinson has been boating all of his life. His work involves teaching spraying courses at PaintTech Training Academy and does some select spraying contract work. He is also the author of a number of books on decorating.

He lives in Preston with his wife Tracey and in the very rare time off likes to relax on his boat.

Check out the website

If you are interested in being kept up to date with future books or you just fancy the odd freebee then subscribe on my website.

www.fastandflawless.co.uk

About the author

Pete Wilkinson has been in the construction industry all of his life. In his younger years he worked for a medium sized decorating company doing a wide range of work.

Then at the age of 27 he got a job teaching Painting and Decorating at a local college. These days he runs his own training academy called PaintTech Training Academy. Why don't you check us out on www.painttechtrainingacademy.co.uk

When I am not working, I like to spend time on my boat with my wife Tracey.

Printed in Great Britain
by Amazon